Mr Hargreaves appeared in his doorway, shouting to the boys to come back. He sounded angry.

'What's up now?' sighed Lenny. 'Looks like we're in for more fireworks than we bargained for this year.'

'Where are my old jeans?' yelled Mr Hargreaves, red in the face. His mood had certainly changed with alarming speed.

'The ones with patches on the knees?'

'My gardening jeans.'

'I thought you didn't need them any more. We used them for the Guy. Mum said we could.'

'You did WHAT?' Mr Hargreaves groaned, before adding in a lowered voice: 'Well, you've done it now, and no mistake! You've wrecked your mother's birthday.'

Also in Beaver by
Hazel Townson

THE LENNY AND JAKE ADVENTURES

FIREWORKS GALORE

Hazel Townson

Illustrated by Philippe Dupasquier

Beaver Books

A Beaver Book

Published by Arrow Books Limited
62-65 Chandos Place, London WC2N 4NW

An imprint of Century Hutchinson Limited

London Melbourne Sydney Auckland
Johannesburg and agencies throughout
the world

First published by Andersen Press 1988
Beaver edition 1989

Printed and bound in Great Britain by
Courier International Ltd, Tiptree, Essex

ISBN 0 09 965540 3

For Anne Marley

Contents

FIREWORKS
GALORE

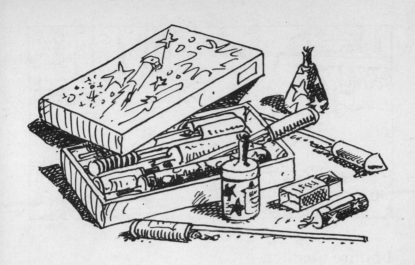

1
A Body Vanishes

If there is one day in the year when the weather really matters it's November Fifth. Lenny Hargreaves leapt out of bed and seized a bunch of curtain. For a minute or two he scrutinised the outside world, then broke into a great grin of relief. For although it was still fairly dark out there, he could tell the day was fine and windless. Just a wisp or two of fog, but nothing to worry about. In fact, perfect

bonfire weather.

From a shelf in the top of his wardrobe Lenny dragged out a huge box of fireworks and sat gloating on the bed. As usual, his mum and dad had turned up trumps, providing the cash for a good selection of fireworks, and his only problem now was how to keep his hands off them until nightfall, when he would be going to the Allens' bonfire.

Lenny's best friend, Jake Allen, belonged to a six-child family which occupied a huge end house with a long back garden. Nobody in the Allen family had the time or the heart for gardening, so the place was a balding wasteland, perfect for building a bonfire. Jake and Lenny had worked hard for weeks now, collecting wood, old furniture, broken toys and other bonfire ingredients, and the result was a miraculous mountain of promise, topped by a truly lifelike Guy—the best in the neighbourhood. What a relief to realise that all the stress and anxiety were nearly over! The great day had come, and all that was left to do now was enjoy it.

Lenny rushed downstairs and grabbed the cornflake packet, shaking his breakfast out with gusto. It would not do to waste a single moment; he must rush round to the Allens' house as soon as possible and say good morning to his creation. For although the garden belonged to the Allens, the bonfire, and in particular the Guy, belonged, Lenny felt, to him. His was the master-mind which had thought up all the places from which to scavenge wood; the means of transport; the method of bonfire-building, and the design of the Guy. Why, the Guy was even wearing Lenny's dad's old jeans and anorak, not to mention his threadbare pyjama-top.

'What's this, then? Last breakfast on the
Titanic?' Lenny's mum enquired as cornflakes
flew in all directions. Before Lenny had time
to think up a smart reply the doorbell rang
and Jake Allen arrived.

Lenny nearly choked. 'What are you doing
here? You're supposed to be guarding the
bonfire!'

'It's okay, my mum's out the back, hanging
washing.'

'Washing? On November Fifth?'

'It'll be dry by tea-time. And if it's not she'll move it into the bathroom, so keep your hair on.' Jake waited until Lenny's mum was out of earshot before whispering, 'Can we borrow the fire-extinguisher off your dad's lorry tonight, just in case?'

Lenny looked incredulous. 'Just in case what? We're supposed to be lighting a fire, not putting one out.'

'Well, ask him, will you?'

'That fire extinguisher belongs to the firm. He's not supposed to move it unless there's a fire.'

'There *will* be a fire, you noodle!'

Lenny regarded his friend suspiciously. 'Has your mum been getting cold feet? She sent you to ask for that fire extinguisher, didn't she?'

Jake looked defiant. 'She says it's a sensible precaution. Our Rod and Gary are only little, and you know what they did with our chip-pan.'

Rod and Gary, Jake's twin brothers, were

certainly full of mischief. Lenny rolled his eyes and sighed.

'Families!'

But as soon as he had said this, Lenny wished he hadn't, for Jake was the eldest of six children and would be lucky if he had more than half a dozen decent fireworks of his own tonight, not that he would ever complain. Jake must think families were a bit of a nuisance too, at times.

'All right, I'll ask my dad,' Lenny relented, 'only he's not back yet.'

Jake began to look worried, for Lenny's dad was a long-distance lorry driver and was often away for days at a time.

'Don't panic, he'll be back before tonight! He never misses Bonfire Night because it's my mum's birthday on the 6th, and he daren't be away for that. Now come on, back to your garden before somebody starts nicking our wood.'

Alas! Had Lenny but known, it was even then too late. A much greater disaster had happened. Their Guy, their wonderful,

stupendous, lifelike Guy, had already
disappeared.

2

A Present Disappears

Lenny's dad arrived home an hour later. His travel-stained lorry drew up at the front gate just as Lenny and Jake trailed miserably round the corner of the street.

'What's up with you two, then?' he called from his cab. 'Your fireworks got dry-rot or something?'

'Hi, Dad!' mumbled Lenny joylessly. 'Somebody's pinched our Guy. We've been all over Cobston looking, but we can't find it.'

'Oh, is that all? Well, there's plenty of time to make another one.' Mr Hargreaves jumped down to the pavement with his bulging zip-bag. 'No use standing about feeling sorry for yourselves. Go and get started.'

'We've no stuff left,' Jake explained. 'We had a sack full of rags for his body, a patched pillowcase for his head, and your old clothes for him to wear.'

Mr Hargreaves laughed. 'Plenty more where that lot came from, I shouldn't wonder. Just let me sort myself out and have a cuppa, then I'll see what I can find you.'

Lenny looked, if anything, even more fed up. 'That's not the point, Dad. We've been made to look daft.'

'Yeah,' agreed Jake, 'whoever's done it, they aren't going to get away with it. We're out for revenge. Only, we've got to find the culprits first.'

'I see your point,' admitted Mr Hargreaves, 'but if I were you I wouldn't let them have the

satisfaction of spoiling my Bonfire Night. I'd make another Guy, even better than the first, then go and enjoy myself and save my revenge for later.'

By now, Mrs Hargreaves had come out to greet her husband. She, at any rate, was all smiles now that Lenny's dad was safely home in time for her birthday. As the two of them disappeared into the house Lenny turned to Jake and declared: 'Never mind what *they* say—honour must be satisfied. Grown-ups

just don't understand.'

'I'll bet it's Bob Langan's lot!' Jake muttered grimly. 'And if it is, I've a good mind to put *him* on our bonfire instead.'

'Jealousy, that's all it is! We've got more wood than anybody round here.'

'We had the best Guy as well.'

Lenny pulled himself together. 'We still have! We're going to find that Guy before dinner-time if it's the last thing we do!' He started dragging Jake away. 'Come on! I know

17

we've had one good look, but we'll just have to have another. The blooming thing must be somewhere.'

The two boys were already halfway down the street when Mr Hargreaves reappeared in his doorway, shouting to them to come back. He sounded angry.

'What's up now?' sighed Lenny. 'Looks like we're in for more fireworks than we bargained for this year.'

'Where are my old jeans?' yelled Mr Hargreaves, red in the face. His mood had certainly changed with alarming speed.

'The ones with patches on the knees?'

'My gardening jeans.'

'I thought you didn't need them any more. We used them for the Guy. Mum said we could.'

'You did WHAT?' Mr Hargreaves groaned and ran a wild hand through his hair, before adding in a lowered voice: 'Well, you've done it now, and no mistake! You've wrecked your mother's birthday. I bought her some ruby earrings and I hid them in the back pocket of

those old jeans because I thought that was the
one place she'd never look. You know how
nosey she is; no good leaving stuff in the
drawers, and if I'd hidden them in one of my
good suits she might have sent it to the
cleaner's whilst my back was turned. I
thought my gardening jeans were safe. Well,
don't just stand there! Go and get 'em back,
double-quick!'

Lenny looked horrified. 'I'm sure there was
nothing in the pockets, Dad. We'd have felt

the corners of a box when we were dressing the Guy.'

'They weren't in a box; just laid on cotton-wool in a little plastic envelope. Only tiny, delicate things they were. Small is beautiful, as your mother always says. Small is expensive, an' all, I can tell you! So you'd better start searching fast!'

'Yeah, okay, Mr Hargreaves,' Jake agreed nervously. 'We'll find them, don't you worry!'

A fine November Fifth this was turning out to be! It couldn't feel any worse being the actual Guy, sitting right on top of the bonfire, watching the first match strike.

3
Quite a Come-back

Bob Langan's Guy had now made an appearance in the garden. Someone had brought it from its secret hiding-place in the garage and propped it against a tree. It was dressed in a tatty old jumper and a pair of running-shorts, and its lop-sided head was made out of a sickly-looking pale green cushion-cover. Certainly it bore no resemblance to Lenny's Guy, which outclassed it by a mile. Peering through the Langans' fence at this monstrosity, Lenny and Jake could not help feeling superior, as well as disappointed.

'I wouldn't be seen dead with a Guy like that!' grinned Jake.

'It's all very well feeling toffee-nosed about it,' Lenny grumbled, 'but at least they've *got* a Guy; we haven't. This was our last hope of finding ours.'

Jake sobered up at once. 'You don't suppose they've just chucked our Guy in the canal?'

'Who knows what folks will do when they're crazed with jealousy?' Lenny thought he had never felt so down in the whole of his life. 'What am I going to tell my dad?'

'Nothing yet. Come round to our house for a bit and we'll have a good think.'

Lenny didn't see what use that would be, but he had no better suggestion. It seemed a long, weary trail back to the Allens' house, and Lenny could not help remembering wistfully the optimism with which he had started his day. It seemed you couldn't be sure of

22

anything in this world. He was jolted from his misery by the sight of Jake's eldest sister Julie running excitedly to meet them.

'Hey, come and look! Our Guy's turned up again!'

'You what?' Jake stood amazed, but Lenny sprinted forward at once.

'Where is it? Who brought it back? I'll have their brains for bike-oil.'

Julie Allen said nobody had seen what happened. She had just found the Guy propped up on the pavement against the back gate, with a great big placard tucked under its arm.

'Placard?' Jake set off after Lenny, fearing the worst.

Well, there it was all right, the Guy they had taken such trouble with. No doubt about its being theirs. Lenny would have known that face anywhere, having himself created every blotch and wrinkle. But what on earth had happened to the Guy? It seemed to have changed its sex and was now wearing a ragged check skirt, a headscarf and a buttonless blouse held together by safety-pins. The plac-

23

ard stuffed under its arm read: UP
WOMEN'S LIB! DOLL FAWKES FOR
EVER!

Lenny was furious. 'Of all the rotten,
sneaky, cheeky . . . !'

'I'll bet I know who's done that,' exploded
Jake. 'Patsy Huggins!'

'Nasty Puggins? I'll pulverise her!' yelled
Lenny, so outraged at the insult that he hadn't
yet realised his dad's jeans were still missing.

'Go on, then!' said a quiet voice behind
him. 'Pulverise me! I'll bet you don't even
know what it means.'

Lenny whipped round, eyes blazing in a
crimson face. And there stood Patsy Huggins,
chin up, feet firmly planted, hands on hips.
Her mouth was wriggling at the corners,
fighting off a grin. Slightly older and taller
than Lenny, she felt confident enough to
stand her ground. Just behind her, though
not looking quite so uppity, was her best
friend, Karen Cook.

For a few speechless seconds Lenny and
Jake stared in startled disgust. Then Jake said:

'You're brainless, Patsy Huggins, you know that? Guy Fawkes was a villain, not a hero. Why do you think we shove him on a bonfire every year? You'll not do Women's Lib. much good with his reputation.'

'Can't take a joke, can you? I might have known it!' Patsy sneered. 'Boys never have a decent sense of humour. Don't you think it's funny?'

'You're going to feel pretty funny,' threatened Lenny, 'if you don't bring that Guy's proper clothes back here in two minutes flat.'

'Oh, keep your hair on, Lenny Hargreaves! We're not thieves,' shouted Karen. 'We haven't stolen your Guy's precious outfit. It's round the front, all in a pile on Jake's doorstep.'

'It had better be!'

The fate of the earrings being top priority, both boys ran off at once to check if this were true, and the girls took the opportunity to slip giggling away.

'You rotten liars!' Jake was soon heard to yell. 'There's nothing on our front doorstep except our Gary's earwig-trapper.' He might as well have saved his breath, for the girls had already disappeared.

4

A Magic Plan

'We did put those clothes back,' Patsy
Huggins declared to an irate Lenny Hargreaves who had followed her home. 'We
folded them up, like we said, and laid them all
neat on one side of Jake's top step.'

'Oh, sure!' sneered Jake. 'And they fell
through a crack in the stone.'

'Honest, it's true!' Karen Cook confirmed
earnestly. 'We only did it for a joke, and we

were going to come along and laugh at you when you started swapping the Guy's clothes back again. We were looking forward to it.'

'Well, where are they, then?'

'Search me!'

'Hey—now I come to think of it,' cried Patsy, 'Bob Langan and his lot were just turning into your street when we left. They'd just come the long way round from their patch, over the canal bridge. Maybe they pinched your stuff.'

'Not them! We've just been looking at their rotten Guy, and it only has shorts on,' Lenny remembered grimly. 'You can see its skinny

old legs made out of grotty sacking tied up with string.'

'Sure, that's what I mean! If ever anybody needed a pair of jeans it's their Guy. They could have nipped back home by the short cut across the playing fields while you were dawdling down here.'

Mad as he was, Lenny could not help feeling that Patsy and Karen were telling the truth. Practical jokes were just their style, but not anything really dishonest. What's more, he could just imagine the Langan gang fetching their Guy out, taking one ashamed look at it and rushing off to find some trousers right away. It was highly likely that they were even now re-clothing their Floppy Fawkes in Lenny's dad's gardening jeans.

'So what are you waiting for?' cried Karen. 'Get round to Bob Langan's, then.'

'We will! And that stuff had better be there, or you two are in dead trouble.' Lenny proceeded to explain about the earrings and the imminent ruin of his mother's birthday. The girls, who had a great respect for birthdays, were thoroughly dismayed.

'Well, if we'd known . . . ' began Patsy.

'We never thought . . .' declared Karen.

'Well, you should try it some time,' Jake said bitterly. 'It's supposed to make your hair grow.'

'They were right, you know!' Jake peered once more through the Langans' fence at the Guy which was now wearing jeans, though so far there was no sign of the other missing garments. Bob and his gang were busy tidying up their bonfire at the other end of the garden and did not spot the spies.

'The utter, brazen cheek of it!' Lenny felt so indignant that he nearly boiled over the top of the Langans' fence like a pan of milk. Jake held him back.

'Simmer down! They're in there, all five of 'em. No use two of us trying to tackle five. We need to use our brains, not our brawn.'

Before Lenny could reply to this, Patsy and Karen stole up behind him. 'Told you it was the Langans, didn't we?'

'I don't know what you're looking so cocky about,' grumbled Jake. 'Those jeans are just about as easy to snatch as the flag on top of Blackpool Tower.'

'Don't tell me you're beaten already!' grinned Patsy. 'Why not get cracking on some of this famous magic of yours, Lenny Hargreaves? Spirit your stuff back through the fence!'

Waving an imaginary wand, she chanted:
'Abracadabra—one—two—three—
Tatty old jeans come back to me!'

'Very funny!' sneered Lenny, who took his magic extremely seriously and intended to make a proper career of it some day. 'But incredible as it may seem, you have actually had a brainwave, Patsy Huggins. And since you two got us into this mess, you can jolly well help to get us out of it again. I *will* do a magic show in Jake's back garden in one hour's time, and you will be my assistants.'

The girls stared at each other in surprise, then Patsy shrugged and said, 'Fair enough! Might be quite a giggle!'

'What do we have to do?' asked Karen, less confidently.

'Just watch he doesn't have you for the Vanishing Accomplice, that's the main thing,' warned Jake, but Lenny ignored him. He was already working out his plans.

'We'll all go back to Jake's and write some tickets out. Then you girls can start giving them away to all the kids in the district, especially to Bob Langan and his lot. Get everybody to write their name on their free ticket and tell them there'll be a prize of half my fireworks for the winning ticket pulled out of a hat.'

'Eh?' gasped Jake. 'Half your fireworks?' He could scarcely believe his ears. Yet Lenny knew that something drastic would have to be done if he was to get his mother's earrings back. Only the chance of a prize would draw the greedy Langans in to an amateur show on an important day like this.

'I'll go and fetch my magic stuff,' Lenny told Jake, 'then Karen can stand at your gate collecting the tickets and Patsy can help me with the tricks.'

'What's Jake going to do? Sell ice-cream at

the interval?'

Lenny gave an exasperated sigh. 'He's going to nip round to the Langans' for the jeans, stupid. That's why we're having a magic show, to get the Langans away from their garden. And we'll make sure the prize draw is the last item of all. Anybody who doesn't stay to the bitter end has no chance of winning. That should give Jake plenty of time.'

Jake was not looking too happy about this.

'Suppose Bob Langan's lot won't come?'

'That's up to these two.' Lenny glared at the girls. 'You just make it sound good and tempting, so they *do* come. Otherwise, I promise you, Jake and me aren't going to forget this for a very long time.'

'Help! We're shivering in our shoes!' grinned Patsy Huggins.

5

A Snake in the Grass

'Free tickets? And a free prize of fireworks? What's the catch?' Bob Langan sounded mighty suspicious.

'No catch,' Patsy Huggins assured him. 'Lenny just needs an audience to practise on before his television audition.'

('I didn't know Lenny was having a television audition,' admitted Karen later, whereupon Patsy said well, he was bound to have one some day, even if he was fifty years old by then. 'He'll never stop pestering, and in the end they'll give him an audition just to shut him up.')

Whether this was true or not, Bob Langan took the bait. He and the other four lads in his gang rushed eagerly off to Lenny's free magic show. In fairness, it must be admitted that this was more in hopes of seeing Lenny Hargreaves make a fool of himself than in hopes of winning a prize, though of course another pile of fireworks was not to be sneezed at.

The gang pushed its way to the front of the crowd, and was ready to start booing and heckling the minute the show began. As soon as Jake had seen Bob's lot settled, he took the short cut to the Langans' house, reconnoitred carefully, then sneaked into the back garden. Jake had to admit he was uneasy. Suppose the Langans got bored with Lenny's crummy tricks and came back home? That lot would tear an intruder to bits before they had even thought about it. He'd better be as quick as possible.

The Guy was still propped against the tree and Jake, sneaking up to it via the privet, was just about to drag the pathetic object into the safety of the empty garage so that he could recapture the earrings, when the back door of the house opened and Bob Langan's mother came out with a bucketful of rubbish for the dustbin.

Jake melted quickly into the shadowy depths of the garage and held his breath. He could see Mrs Langan through the gap in the door, and watched with galloping heart as she replaced the dustbin lid, then picked up a broom and started sweeping the path! Oh, boy! At this rate he could be stuck in the garage for half an hour! How long could Lenny keep going?

'It's all right for some!' Jake thought, aggrieved. Here was he, risking life and limb, while Lenny Hargreaves messed about pulling flags out of cocoa-tins and wriggling crumpled paper flowers down his sleeve. Next time, Lenny could do his own dirty work.

Whilst Jake was quietly fuming, Mrs Langan noticed that the garage door was ajar, came over to close it properly, and actually locked it! Jake heard the key turn in the lock, and it might as well have been the last nail going into the lid of his coffin. He was done for!

'As you can see,' declared Lenny Hargreaves, holding aloft a metal cylinder, 'this container is completely empty. But in case you don't trust me, perhaps one member of the audience would like to volunteer to check inside it?'

With a great crow of delight, Bob Langan leapt to Lenny's side and rammed his fist into the cylinder. Then with an agonised shriek he started capering about as if in pain, pretending he couldn't get the cylinder off his arm.

'Ow! Help! I'm crippled for life!' he yelled, bouncing up and down in front of Lenny's magic table, setting the objects on it dithering. But Lenny, undismayed, tapped the back of Bob's neck with his magic baton and produced a small but lively snake from inside Bob's shirt collar. The audience roared with laughter, while Patsy Huggins brought her foot down with a powerful stamp onto Bob Langan's toes, at the same time seizing the cylinder and wrenching it free.

Bob Langan was half furious, half scared. He couldn't stand snakes at any price. Was this one real or not? Magicians' rabbits and pigeons were real, weren't they? Bob swung away in panic and crashed into Patsy, forcing her backwards into the magic table. Patsy lost

her balance and upset a great, black bagful of feathers all over Bob. This brought the rest of the Langan gang to their feet, and a spectacular scrum was soon under way, with feathers, coins and playing cards flying in all directions.

'Watch out for my snake!' yelled Lenny at the top of his voice. 'It bites if it gets scared!' And suddenly the Langan gang was gone, legging it down the road like Olympic sprinters. Shrieking and scuffling, the rest of the audience followed.

Meantime, Jake Allen, in growing alarm, was still struggling to find a way out of the Langans' garage. There was a small window high up on the left, and he had managed to drag a workbench to a spot underneath it, so that he could reach, although he already suspected the window was too small for him. He now measured the space with his hands and found himself to be half a hand wider than the window.

Well, what about mind over matter? If he

took deep breaths, like the yogi were supposed to do, would he be able to squeeze through the gap? (Surely that was easier than lying on a bed of nails or being buried alive, which yogi seemed to manage with no trouble at all.) But no; Jake could see he would only get stuck halfway—a fate too horrible to contemplate.

Jake was fast running out of ideas when he heard the gate bang open and footsteps come thumping towards the garage. The Langans were back! Desperate, he peered anxiously out of one corner of the window—and saw not the Langans, but Karen Cook staring wildly round the garden!

Saved! Jake banged on the glass to attract Karen's attention, then expertly mimed his fate like the good little actor he was. Karen immediately dashed round to the garage door, found the key in the outside lock and set Jake free, breathlessly describing the undignified end to the magic show.

'When I saw what was happening I thought I'd better warn you Bob Langan's lot were coming back. I'm only just ahead of them.'

So there was no time for stripping off the Guy's jeans, or even searching its pockets. Jake picked the whole Guy up and ran.

'Go the long way round, over the canal bridge, then they won't see us!' yelled Karen.

But Jake knew that hostilities were only a matter of time. His theft of the Guy was as good as a declaration of war.

6

Two Swallows Cause a Panic

Lenny Hargreaves carefully tidied away his magic tricks. This was an important task, because if it was not properly done the tricks might not work next time. Every silk scarf, every playing card, every jar and box and lid had to be set correctly in its place, and the silk snake with the special spring inside it had to be fitted back into the hollow end of his baton. Quite undaunted by the chaotic finish

47

to his show, Lenny felt pleased that he had played for time and won. Jake must have found the earrings by now, and as an added bonus, the show had broken up in such confusion that no one had stayed to the end for the prize draw. So all Lenny's fireworks were still intact. As if that were not enough, he had actually managed to impress Patsy Huggins.

'Hey, that snake trick wasn't bad!'

'Oh, nothing to it! That's just one of the simple ones. You should see the "Sawing Your Thumb Off" or the "Flying Floormop".'

'How do you learn it all?'

'Well, you have to have a special flair for it,' boasted Lenny. 'Nimble fingers and good co-ordination and a quick eye and that. Magicians are born, not made.'

'Oooh, hark at you!' Patsy had not been all *that* impressed. 'How come that tenpence fell down your sleeve when it shouldn't have, then? And how come those metal rings weren't joined up like you said they would be?'

Lenny was saved from replying by the timely appearance of Jake and Karen with the Langans' Guy.

With a triumphant cry, Lenny plunged his hands into first one pocket, then another of the Guy's jeans before beginning to look distinctly sick.

'Idiot! You've gone and lost the earrings on the way back!'

'No, I haven't! We took special care, didn't we, Karen?'

'Where are they, then?'

'Aren't they there? You mean to say I've gone through all that for nothing?' Jake was outraged. 'I could have been stuck in that garage for weeks. The Langans haven't even got a car.'

'I'll bet Bob Langan found your earrings first. He'll have sold them to a jeweller for money to buy more fireworks,' said Karen.

Lenny groaned. This was the absolute, bitter end. He sat down on a lump of wood and felt distinctly suicidal.

And then he noticed something.

'Hey—those aren't my dad's jeans! They had bigger patches than that. They were his gardening jeans, and even the patches had patches. My mum made them specially thick, like kneeling mats.'

'So we're at war with the Langans all for nothing?' asked Jake in disgust, while Karen asked, 'What happened to the proper jeans, then?'

'Simple! We left them on Jake's step, so I expect his mum came out and picked them up,' explained Patsy.

'Oh, no!' groaned Jake. 'While we've been messing about, my mum's probably cut those jeans up and made two pairs for our Rod and Gary.' He shot feverishly into the house to find out if this were so, but his mother was not there. She had just rushed off to the doctor's with Rod and Gary, Julie said.

'Why, what's up with 'em?'

Julie shrugged. 'Seems they said they'd eaten some shiny red berries. Mum thought they might be poisonous.'

'Shiny red berries?' echoed Lenny

Hargreaves. 'Well, you know what those are, don't you?'

'Hips and haws?' suggested Karen.

'Holly,' said Patsy.

'Deadly nightshade!' declared Jake.

'My mum's earrings,' corrected Lenny grimly. 'They must have had one each.'

7
Don't Lose Your Head!

'Right!' said Lenny bossily. 'This is where we split up again. Somebody run down to the doctor's to sort out the truth about these so-called berries, and somebody go and put the Langans' Guy back.'

'Don't look at me!' warned Jake. 'I'm not going anywhere near the Langans' again.'

'Chicken!' Lenny grabbed the Langans' Guy and hoisted it on to his shoulder, where-

upon its head fell off.

'These two are playing games with you,' the doctor told Mrs Allen. 'The only thing they've swallowed recently is raspberry jam.'

Rod and Gary grinned furtively at one another, while Mrs Allen made angry, threatening noises.

'There was berries in the jam,' protested Rod.

'Big, fat ones,' agreed Gary.

Seizing one twin in each hand, Mrs Allen dragged them roughly from the surgery.

'So this was your idea of a joke? Just wait till

I get you home! There'll be no bonfire for you two, that's for certain!'

'Oh, Mum! That's not fair!'

'Specially when we found you all those nice new clothes.'

'What nice new clothes? What tale are you making up now? I tell you this, if you two don't stop telling such whopping lies . . . .'

'It's not lies!'

'No, it's not! We found them on your step.'

'We thought you'd be pleased. You're always saying you've got no money for new things.'

'We put them on your bed.'

Just then Jake ran round the corner and nearly bowled his mother over.

'Where are Lenny's earrings?' he yelled. 'I'll murder you two!'

'What's Lenny Hargreaves doing with earrings, may I ask? Has he joined a pirate ship, or what?' Long before Mrs Allen had sorted all this out, she was home again and staring in disgust at the pile of strange garments that had found its way on to her nice, clean bedspread.

'Straight on to the bonfire with that lot! Must be crawling with germs.'

'Hang about, Mum! Don't you ever listen?' pleaded Jake. 'All the way home I've been trying to tell you about Lenny's earrings. Oh, what's the use?' Snatching the jeans from the pile of clothes, he turned them upside down and shook them vigorously. But once again there was nothing in the pockets.

'Give that head to me,' said Patsy Huggins. 'I'll stitch it back on again and make a better job of it than the Langans ever did.'

Lenny handed over the sickly green object. He had lost interest completely.

'It's all your fault, Patsy Huggins!' cried Jake. 'If you hadn't stripped our Guy in the first place, none of this would have happened.'

'Well, I can't go on saying I'm sorry till Doomsday,' retorted Patsy. 'The best I can do is go and own up to Lenny's dad.'

'Can't say fairer than that,' agreed Karen.

'I'll go, too.'

'That won't bring the earrings back,' Lenny pointed out.

'Well, it's not the end of the world. He'll just have to buy your mum something else. The shops don't close until six.'

'What a rotten, horrible day!' moaned Jake.

'Honest, you two are a right pair of miseries!' Patsy burst out. 'Why can't you look on the bright side of things for a change? If your dad buys your mum a big box of chocolates instead of earrings, you and Jake will probably get to eat half of them yourselves. You don't know when you're lucky!'

'Oh, come on! Let's get it over with!' Lenny tossed the headless Guy into Jake's shed and began to lead the way home, just as the Langan gang started arming itself for the kill.

8
Kidnap and Collision

'I can't believe it!' Mr Hargreaves cried. He plunged a hand into his bedside cupboard and gathered up a small plastic envelope in which a pair of ruby earrings winked and sparkled on a bed of cotton wool.

'I could have sworn I put them in the pocket of my gardening jeans.' Then he remembered that he had moved his jeans into the cupboard just before he went away, to make extra sure the earrings were not found. Somehow the package must have fallen out unnoticed when the jeans were taken for Lenny's Guy.

Once he had recovered from his shock, Mr Hargreaves' next thought was for Lenny. The poor lad must be frantic, searching for something he would never find. Better seek him out and give him the glad tidings at once.

'Just going out to look for Lenny,' Mr Hargreaves called to his wife, before jumping into his lorry and driving off in the direction of the Allens'.

When Lenny and his cronies arrived home, Mrs Hargreaves told them her husband was out looking for Lenny.

'Must be important, too, for he took the lorry.'

Lenny's heart sank into his socks. He knew perfectly well why his dad was looking for him, though of course he could not tell his mum. So the frustrated little party turned away.

'May as well go and sew that head on and

change your Guy back into his proper clothes,' said Patsy.

Karen Cook's face suddenly lit up. 'Hey— I've just had an idea!' She began to whisper in Patsy's ear, but all Jake could say was, 'Haven't you two had enough ideas for one day?'

'I'll tell you *my* idea. I've a blooming good mind to run away to sea,' said Lenny. 'As a matter of fact, I've often fancied being a pirate. *Making* trouble, instead of just dropping into it.'

Bob Langan's gang turned up at Jake's house armed to the eyebrows with staves of bonfire wood and catapults and a wooden rolling-pin from Mrs Langan's kitchen. They rushed the back gate and burst into the garden, to find no one there but Rod and Gary, playing football with the sickly green head of a pathetic Guy Fawkes.

Bob Langan recognised the head and nearly had a fit. He seized young Gary and carried him kicking and struggling into

the street.

'Grab the other one as well!' he yelled. 'We'll teach 'em to take liberties with our property.'

But Rod was too quick for them. He had already darted into the house and slammed the back door in their faces. They were forced to leave without him as they heard him yelling up the stairs:

'Mum! Some big boys have kidnapped our Gary!'

'Are you at it again, telling your lies? I honestly don't know how you dare, our Rod!

Well, this is one time too many, and I wouldn't like to be you when your dad comes home.'

'It's true, Mum! It was that big lad with the snake down his neck this morning.'

'You're sure he wasn't riding on an elephant as well?' sneered Mrs Allen, getting on with her bed-making. If only she had looked out of the window, she might have seen Patsy Huggins sneaking into the garden and gathering up a battered green head.

It happened in a flash! There was a shout and a screech of brakes as the lorry swung across the road and slithered to a halt, missing the children by inches. They seemed to have sprung from nowhere, first a big boy holding a smaller boy in his arms, then four other boys brandishing a motley assortment of weapons. The big boy fell; the small boy tumbled from the big boy's arms, staggered to his feet and ran yelling towards the lorry.

'Mr Hargreaves! They're trying to kidnap me!'

Lenny's dad jumped down from his cab and gathered up young Gary, who did not seem to be hurt. He was quite upset, though, and by the time Mr Hargreaves had calmed him down the Langan gang had disappeared.

Lenny's dad took young Gary straight home and Mrs Allen, suddenly full of concern, put the kettle on.

'I thought our Rod was just telling another silly tale,' she confessed. 'I'll never forgive myself! If you hadn't rescued Gary, I dread to think what would have happened.'

Mr Hargreaves smiled. 'Oh, I don't expect they would have done anything drastic. They just wanted to scare him a bit in revenge for their ruined Guy.' He could not help remembering that his own son and Jake had expressed just such feelings that very morning, when their own Guy was lost.

'Yes, well, if our Jake's had something to do with this, I shall have a few words to say to him when he comes in. What with one thing and another, I don't think they deserve a bonfire tonight, any of them.'

'Oh, let's be fair,' said Mr Hargreaves. 'I think it's really all my fault. I seem to have started a big wild-goose chase.'

He began to explain about the missing earrings.

Just as he was finishing his tale, Lenny and Jake came in. Lenny stopped and turned pale when he spotted his dad, and would have made for the door again, but Mr Hargreaves hastened to explain.

'Just a minute, lads! Your troubles are over!'

The dawning relief on the boys' faces was proof enough that they had really suffered, and Mr Hargreaves was determined to make it up to them.

'Tell you what, this year I think you deserve the best Bonfire Night you've ever had. We can manage a few more fireworks, for a start. Then some treacle toffee and chestnuts. And how about pie-and-chips all round when everyone's got nice and hungry?'

Lenny's eyes lit up like sparklers.

'Hey, Dad, that'd be great!'

'Can we stay up really late?' asked Jake, snatching at this rare opportunity of adult benevolence. 'And can we nip round to the Langans' as well, to see how their fire's getting on?'

'I thought you and the Langans were sworn enemies?'

Lenny grinned. 'We want to see their Guy, Dad. According to Patsy Huggins it's supposed to be setting a new trend.'

'Yeah,' Jake chuckled, 'something to do with Women's Lib.'

HAZEL TOWNSON

If you're an eager Beaver reader, perhaps you ought to try some of our exciting and funny adventures by Hazel Townson. They are available in bookshops or they can be ordered directly from us. Just complete the form below and enclose the right amount of money and the books will be sent to you at home.

☐ THE SPECKLED PANIC	£1.50
☐ THE CHOKING PERIL	£1.25
☐ THE SHRIEKING FACE	£1.50
☐ THE BARLEY SUGAR GHOSTS	£1.50
☐ DANNY—DON'T JUMP!	£1.25
☐ PILKIE'S PROGRESS	£1.95
☐ ONE GREEN BOTTLE	£1.50
☐ GARY WHO?	£1.50
☐ THE GREAT ICE-CREAM CRIME	£1.50
☐ THE SIEGE OF COBB STREET SCHOOL	£1.25

If you would like to order books, please send this form, and the money due to:
ARROW BOOKS, BOOKSERVICE BY POST, PO BOX 29, DOUGLAS, ISLE OF MAN, BRITISH ISLES. Please enclose a cheque or postal order made out to Arrow Books Ltd for the amount due including 22p per book for postage and packing both for orders within the UK and for overseas orders.

NAME . . . .

ADDRESS . . . .

. . . .

Please print clearly.

Whilst every effort is made to keep prices low it is sometimes necessary to increase cover prices at short notice. Arrow Books reserve the right to show new retail prices on covers which may differ from those previously advertised in the text or elsewhere.

JOAN LINGARD

If you enjoyed this book, perhaps you ought to try some of our Joan Lingard titles. They are available in bookshops or they can be ordered directly from us. Just complete the form below and enclose the right amount of money and the book will be sent to you at home.

☐ MAGGIE 1: THE CLEARANCE	£1.99
☐ MAGGIE 2: THE RESETTLING	£1.99
☐ MAGGIE 3: THE PILGRIMAGE	£1.95
☐ MAGGIE 4: THE REUNION	£1.95
☐ THE FILE ON FRAULEIN BERG	£1.99
☐ THE WINTER VISITOR	£1.99
☐ STRANGERS IN THE HOUSE	£1.95
☐ THE GOOSEBERRY	£1.95

If you would like to order books, please send this form, and the money due to:
ARROW BOOKS, BOOKSERVICE BY POST, PO BOX 29, DOUGLAS, ISLE OF MAN, BRITISH ISLES. Please enclose a cheque or postal order made out to Arrow Books Ltd for the amount due including 22p per book for postage and packing both for orders within the UK and for overseas orders.

NAME . . . .

ADDRESS . . . .

. . . .

Please print clearly.

Whilst every effort is made to keep prices low it is sometimes necessary to increase cover prices at short notice. Arrow Books reserve the right to show new retail prices on covers which may differ from those previously advertised in the text or elsewhere.

BEAVER BOOKS FOR YOUNGER READERS

Have you heard about all the exciting stories available in Beaver? You can buy them in bookstores or they can be ordered directly from us. Just complete the form below and send the right amount of money and the books will be sent to you at home.

☐ THE BIRTHDAY KITTEN	Enid Blyton	£1.50
☐ THE WISHING CHAIR AGAIN	Enid Blyton	£1.99
☐ BEWITCHED BY THE BRAIN SHARPENERS	Philip Curtis	£1.75
☐ SOMETHING NEW FOR A BEAR TO DO	Shirley Isherwood	£1.95
☐ REBECCA'S WORLD	Terry Nation	£1.99
☐ CONRAD	Christine Nostlinger	£1.50
☐ FENELLA FANG	Ritchie Perry	£1.95
☐ MRS PEPPERPOT'S OUTING	Alf Prøysen	£1.99
☐ THE WORST KIDS IN THE WORLD	Barbara Robinson	£1.75
☐ THE MIDNIGHT KITTENS	Dodie Smith	£1.75
☐ ONE GREEN BOTTLE	Hazel Townson	£1.50
☐ THE VANISHING GRAN	Hazel Townson	£1.50
☐ THE GINGERBREAD MAN	Elizabeth Walker	£1.50
☐ BOGWOPPIT	Ursula Moray Williams	£1.95

If you would like to order books, please send this form, and the money due to:
ARROW BOOKS, BOOKSERVICE BY POST, PO BOX 29, DOUGLAS, ISLE OF MAN, BRITISH ISLES. Please enclose a cheque or postal order made out to Arrow Books Ltd for the amount due including 22p per book for postage and packing both for orders within the UK and for overseas orders.

NAME ....

ADDRESS ....

....

Please print clearly.

Whilst every effort is made to keep prices low it is sometimes necessary to increase cover prices at short notice. Arrow Books reserve the right to show new retail prices on covers which may differ from those previously advertised in the text or elsewhere.

BEAVER BESTSELLERS

You'll find books for everyone to enjoy from Beaver's bestselling range—there are hilarious joke books, gripping reads, wonderful stories, exciting poems and fun activity books. They are available in bookshops or they can be ordered directly from us. Just complete the form below and send the right amount of money and the books will be sent to you at home.

☐ THE ADVENTURES OF KING ROLLO	David McKee	£2.50
☐ MR PINK-WHISTLE STORIES	Enid Blyton	£1.95
☐ FOLK OF THE FARAWAY TREE	Enid Blyton	£1.99
☐ REDWALL	Brian Jacques	£2.95
☐ STRANGERS IN THE HOUSE	Joan Lingard	£1.95
☐ THE RAM OF SWEETRIVER	Colin Dann	£2.50
☐ BAD BOYES	Jim and Duncan Eldridge	£1.95
☐ ANIMAL VERSE	Raymond Wilson	£1.99
☐ A JUMBLE OF JUNGLY JOKES	John Hegarty	£1.50
☐ THE RETURN OF THE ELEPHANT JOKE BOOK	Katie Wales	£1.50
☐ THE REVENGE OF THE BRAIN SHARPENERS	Philip Curtis	£1.50
☐ THE RUNAWAYS	Ruth Thomas	£1.99
☐ EAST OF MIDNIGHT	Tanith Lee	£1.99
☐ THE BARLEY SUGAR GHOST	Hazel Townson	£1.50
☐ CRAZY COOKING	Juliet Bawden	£2.25

If you would like to order books, please send this form, and the money due to:

ARROW BOOKS, BOOKSERVICE BY POST, PO BOX 29, DOUGLAS, ISLE OF MAN, BRITISH ISLES. Please enclose a cheque or postal order made out to Arrow Books Ltd for the amount due including 22p per book for postage and packing both for orders within the UK and for overseas orders.

NAME . . . .

ADDRESS . . . .

. . . .

Please print clearly.

Whilst every effort is made to keep prices low it is sometimes necessary to increase cover prices at short notice. Arrow Books reserve the right to show new retail prices on covers which may differ from those previously advertised in the text or elsewhere.

COLIN DANN

Colin Dann's exciting animal stories are available in Beaver. You can buy them in bookshops or they can be ordered directly from us. Just complete the form below and send the right amount of money and the books will be delivered to you at home.

☐ IN THE GRIP OF WINTER	£2.50
☐ FOX'S FEUD	£1.99
☐ THE FOX CUB BOLD	£1.99
☐ THE SIEGE OF WHITE DEER PARK	£1.99
☐ THE RAM OF SWEET RIVER	£2.50
☐ THE KING OF THE VAGABONDS	£1.99

JACK LONDON

The classical animal adventure stories by Jack London are thrilling books telling of the lives of the mane and animals in Alaska during the great Gold Rush. They can also be bought in bookshops or ordered from us.

☐ WHITE FANG	£1.95
☐ CALL OF THE WILD	£1.50

If you would like to order books, please send this form, and the money due to:
ARROW BOOKS, BOOKSERVICE BY POST, PO BOX 29, DOUGLAS, ISLE OF MAN, BRITISH ISLES. Please enclose a cheque or postal order made out to Arrow Books Ltd for the amount due including 22p per book for postage and packing both for orders within the UK and for overseas orders.

NAME ....

ADDRESS ....

....

Please print clearly.

ENID BLYTON

If you are an eager Beaver reader, perhaps you ought to try some of our exciting Enid Blyton titles. They are available in bookshops or they can be ordered directly from us. Just complete the form below, enclose the right amount of money and the books will be sent to you at home.

☐	THE CHILDREN OF CHERRY-TREE FARM	£1.99
☐	THE CHILDREN OF WILLOW FARM	£1.99
☐	NAUGHTY AMELIA JANE	£1.50
☐	AMELIA JANE AGAIN	£1.50
☐	THE BIRTHDAY KITTEN	£1.50
☐	THE VERY BIG SECRET	£1.50
☐	THE ADVENTUROUS FOUR	£1.50
☐	THE ADVENTUROUS FOUR AGAIN	£1.50
☐	THE NAUGHTIEST GIRL IS A MONITOR	£1.99
☐	THE NAUGHTIEST GIRL IN THE SCHOOL	£1.99
☐	THE ENCHANTED WOOD	£1.99
☐	THE WISHING-CHAIR AGAIN	£1.99
☐	HURRAH FOR THE CIRCUS	£1.75

If you would like to order books, please send this form, and the money due to:
ARROW BOOKS, BOOKSERVICE BY POST, PO BOX 29, DOUGLAS, ISLE OF MAN, BRITISH ISLES. Please enclose a cheque or postal order made out to Arrow Books Ltd for the amount due including 22p per book for postage and packing both for orders within the UK and for overseas orders.

NAME ....

ADDRESS ....

....

Please print clearly.

Whilst every effort is made to keep prices low it is sometimes necessary to increase cover prices at short notice. Arrow Books reserve the right to show new retail prices on covers which may differ from those previously advertised in the text or elsewhere.